This 8 hp lamp-start oil engine built in 1920 by Keighley Gas & Oil Engine Company Limited spent its working life driving a corn crusher on a farm near Scarborough.

Old Stationary Engines

David W. Edgington

Published by Shire Publications Ltd,
PO Box 883, Oxford, OX1 9PL, UK
PO Box 3985, New York, NY 10185-3985, USA
Email: shire@shirebooks.co.uk   www.shirebooks.co.uk

© 1980 by David W. Edgington.

First published 1980
Transferred to digital print on demand 2014.

Shire Library 49.
ISBN: 978 0 74780 594 6.
David W. Edgington is hereby identified as the author of this
work in accordance with Section 77 of the Copyright, Designs
and Patents Act 1988.

British Library Cataloguing in Publication Data:
Edgington, D. W. (David William)
Old stationary engines. – 2nd ed. – (Shire Library; 49)
I. Title 621.1'6
ISBN: 978 0 74780 594 6

COVER IMAGE: *The Lister A type, built between 1923 and 1963 was extremely popular, and built in large numbers. This example, built to specification 15 and one of the last with caged valves, left the factory in January 1926.*

ACKNOWLEDGEMENTS
Illustrations are acknowledged as follows: Ole Baekkedal, page 51 (bottom); Andrew Borman, page 20 (top); Eric Brain, front cover, pages 36 (right), 39 (both), 42 (top), 47 (bottom), 49 (top), 50 (bottom); John Burning, page 48 (top); Jamie Coates, page 35 (both); Iwan Evans, page 26 (lower left); Len Gillings, page 44 (bottom); George Hayton, pages 19 (bottom), 21 (bottom), 33 (centre and bottom), 50 (top); David Howson, page 45 (bottom); J. L. A. Kammer, page 25 (top); Peter Keeley, page 45 (top); Mark Kinzie, page 52 (top); Garry Montgomery, page 43 (top); S. Morris, page 37 (top); John Murray, page 24 (bottom); Ted Neale, page 44 (centre); M. C. Randall, page 11; Arnold Sayer, page 1; Ken and Sean Sparkes, page 28 (top); Jim Taylor, page 27; Philip Thornton-Evison, pages 23, 26 (bottom right), 29 (both), 30 (top and right), 31 (top), 37 (centre), 38 (centre), 41 (all), 42 (centre), 47 (top and right), 54. Other illustrations are from the author's collection or are photographs by the author.

Printed and bound in Great Britain.

MIX
Paper from
responsible sources
FSC® C013604

Shire Publications is supporting the Woodland Trust, the UK's leading woodland conservation charity, by funding the dedication of trees.

Contents

A stationary engine is an engine that delivers its power from a fixed position. Here a Bentall engine is driving barn machinery via line-shafting, 1911.

The development of the stationary engine

The earliest attempt to build an internal combustion engine was probably the occasion in 1673 on which a famous Dutch physicist and astronomer, Christiaan Huygens, constructed a vertical engine that consisted of a cylinder in which a piston was caused to ascend by exploding a charge of gunpowder. Needless to say, the charge was not successful because after each stroke the base of the cylinder required removal in order to insert a fresh charge. A comparison can be made between this engine and a gun, because the latter could be termed a one-stroke internal combustion engine that throws away its piston at each stroke.

About two hundred years passed before French and German innovators finally came near to producing a prime mover that was destined to become cheaper to manufacture and operate than the 'external combustion' steam engine. The various ideas promoted by these very early inventors concerning this class of engine are numerous. Not only are they exceedingly difficult to classify because of their diversity, but no one has yet been able to weed out incorrect or erroneous notions from among so many that are either not fully recorded or even untried.

It was the introduction of coal gas that led inventors to the possibilities of using a mixture of this gas with air for the generation of mechanical power. A Frenchman, Etienne Lenoir, was the first person to achieve any degree of success with an operational gas engine. His merit lies not in his having been the first to think out the details of a gas engine but in the fact that

The earliest vertical engines adopted the overhead crankshaft style, with piston pointing downwards, as used by contemporary steam engines and steam pumps. Nowadays this somewhat incongruous design tends to be referred to as 'inverted vertical' by engine enthusiasts. The Midland gas engine seen here was constructed by Nottingham engineers John Taylor & Sons and first appeared at the Doncaster Show in 1891. Two cylinders are used; one is a working cylinder while the other is a pump in which the charge of gas and air is mixed. Tube ignition is used to fire the charge. Output was claimed to be a surprisingly high 3½ hp.

he made one and it worked. His engine (1860) used the double-acting steam-engine principle to inject coal gas first into one end of the cylinder and then into the other. When the piston was halfway along the cylinder, the gas and air mixture was lit by means of a spark. It should be noted that the gas and air mixture was not compressed in the cylinder before burning; in other words, a non-recoiling action was used. Engines that did not compress their mixture before firing it worked uneconomically and soon had to give way to more modern constructions.

The major breakthrough came in 1876 when N. A. Otto produced an engine that operated by using the familiar four-stroke or 'Otto' cycle that is so well known nowadays. His engine also used coal gas as fuel, ran twice as fast as Lenoir's engine, yet used only half as much fuel. The advantage achieved by Otto was obtained by compressing the gas and air mixture before combustion, so adding enormously to economy

The NEW "OTTO" SILENT GAS ENGINE.

Cost of Gas, 1d. per Hour ;per indicated H.-P., at 4s. per 1,000 feet.

Sizes at present offered.

Nominal. Actual.

H.-P.		H.-P.
1	=	2
2	=	3
3½	=	5
8	=	12
16	=	30

This Engine is made by the Makers of the Atmospheric Gas Engine, of which nearly 5,000 have been sold.

THE NEW SILENT ENGINE unites the greatest simplicity of parts ever yet attained in a Gas Engine, or even in many Steam Engines, with an economy and durability often surpassing either. It is as silent as a Steam Engine, and of course has the immense advantage of starting at full power on the gas being lit; and, by dispensing with the Boiler, of avoiding all risks, annoyances, and attention, which a Boiler entails. The cost of replacing Steam Power by Gas has in some cases been recovered in less than two years, being a saving of over 50 per cent. per annum on the outlay. Prices, &c., on application to

CROSSLEY BROTHERS { Works - Great Marlborough-street, Manchester.
London Warehouse—116, Queen Victoria-street, E.C.
Sole English and Colonial Makers.

The Otto Silent gas engine introduced the world to the four-stroke or 'Otto' cycle when it arrived in 1876. Here was an engine with an immediate capability to usurp the inefficient and costly steam engine. Its method of ignition was by carrier flame – a mechanically operated slide valve carried the flame to the cylinder to ignite the charge. This '5 man-power' version was advertised by Crossley Brothers in 1879.

Joseph Day of Bath pioneered the two- and three-port two-stroke system. Before 1900 there was a tendency to build multi-cylinder engines separately and then link them together, as shown here with Day's Twin gas engine of 1892.

and fuel efficiency. This great advance in technology was an important stepping-stone in the evolution of the modern internal combustion engine. Here was the first practical stationary engine, which proved suitable for use in the small workshop, factory or mill. It soon found a wide application in industrial establishments for the generation of comparatively small powers. It provided at last a source of power cheap enough for businesses that could not afford to buy and fuel the expensive steam engine.

The pioneers of the petroleum industry, in producing several brands of paraffin for illumination purposes, provided the key to the birth of the 'liquid' fuel-burning engine. Initially, from about 1880, highly volatile fuels were tried, these being similar to what is now termed 'petrol' but often even more volatile. However, the dangers of storing these fuels and adapting them for commercial use forced pioneers to look for a much safer substitute. The first to achieve success in this direction were the brothers William and Samuel Priestman, who began experimenting with lamp oil as fuel in the early 1880s.

Paraffin (lamp oil) requires heat treatment in order to vaporise it before it will form a suitable mixture (with air) for combustion. With the Priestman engine, the fuel was delivered to the cylinder in an atomised condition by pressurising the fuel tank, from where the paraffin was passed through a sprayer to an exhaust-heated chamber into which additional air was delivered on the suction stroke of the piston. Although a blowlamp is required to heat the vaporising chamber when starting the engine, once it is running the charge is fired by electric ignition. This type of engine worked on the Otto cycle and compressed the vaporised mixture before the point of ignition.

During the 1890s two significant steps were taken in the evolution of the oil engine. The engines mentioned so far were all designed for the induction and compression of a pre-mixed charge of gas or oil with air. The extent to which an explosive mixture can be compressed is limited, so the next step in oil-engine genealogy is a salient one. In 1890 Herbert Akroyd Stuart introduced two important features, the first of which was the idea of vaporising and igniting the fuel by bringing it into contact with the hot wall of the combustion chamber at a pre-arranged time, thus timing the actual point of ignition by injecting fuel at exactly the right moment. The second notable feature of the Akroyd engine was that only pure air was drawn into the cylinder and compressed – no ignition being possible until the fuel was injected into the bulb-shaped projection situated at the end of the cylinder. For starting this type of engine a blowlamp is used to heat the bulb (termed, for obvious reasons, a 'hot-bulb'), but once the engine is running the

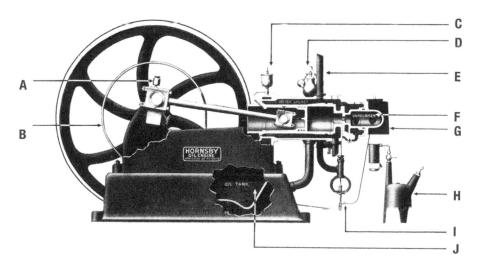

Hornsby oil engine: section through cylinder. The hot-bulb can be seen at the end of the cylinder and is marked 'vapouriser' [sic]. The starting blowlamp is positioned under it. Key: A, cotton-waste lubricator (the cotton acts as a siphon and feeds the bearing by capillary attraction); B, crank and splash-guard, a necessary feature on any open crankshaft engine; C, sight feed lubricator for oiling the piston and small end bearing; D, centrifugal governor; E, cooling water outlet pipe; F, vaporiser; G, vaporiser heat shield; H, starting lamp; I, fuel pump; J, fuel tank situated in engine base.

blowlamp can generally be dispensed with as ignition is maintained by compression, aided by heat retained in the hot-bulb after each explosion. Akroyd Stuart's engine was manufactured by R. Hornsby & Sons Limited and was known as the Hornsby-Akroyd. It was the first successful hot-bulb engine. The principle of injecting the fuel into the hot-bulb vaporiser was soon adopted in other countries, including Germany, Sweden, Norway, Denmark and Russia.

Rudolph Diesel designed an engine that compresses pure air to such an extent that it becomes hotter than the ignition point of the fuel. Hence, if a suitable fuel is mixed with air it 'explodes' from the heat without any need for sparks from a magneto or pre-heating with a blowlamp. As with the Akroyd engine, the fuel was injected just at the required time, but with the Diesel system the highly compressed air in the cylinder ignited the fuel, whereas with the Akroyd engine it was the heat retained in the hot-bulb, aided by compression (which was much lower), that produced the ignition. Although the engines of Rudolph Diesel and Herbert Akroyd Stuart are totally different in principle and construction, both were the forerunners of the much-developed lightweight high-speed engines in use today.

The evolution of the early stationary engine in its common forms has been briefly described, but there is one further style to which reference must be made. This is the two-stroke, in which there is an explosion every revolution instead of every other revolution as with the Otto or four-stroke cycle. The same process takes place as with the four-stroke engine – suction, compression, explosion, exhaust and so on – but instead of each operation taking a full stroke, they are all carried out in two strokes – or one revolution of the engine.

Robson, Clerk, Day, Nash (in the United States), Fielding and quite a number of others first applied the two-stroke cycle in its various forms, all using their own particular style. As stated, there were many variations on this theme and only a brief classification of each can be given here: Robson – under piston scavenge; Clerk – scavenge pump; Day – three port; Nash – two port; Fielding – uniflow. There are many contemporary technical books on this subject.

From about 1900 stationary engines were being produced in their thousands and virtually every small town had an engine manufacturer of some description. Whether it was the agricultural blacksmith, the students with a project at the college, merely a do-it-yourself engineer with a garden workshop or the small factory designed solely for the production of gas and oil engines, all played a significant part in the exciting evolution of the internal combustion engine. These new mechanical workhorses were finding numerous applications, of which the most popular were providing even the smallest workshop or factory with an independent source of power. Stationary engines drove generators for household lighting and were used on almost every farm for driving machinery for grinding corn, pumping water, sawing wood, cake crushing and many other tasks. The main advantages of this form of power-plant lay in its portability (if required), economy of running, its reliability and the fact that – providing fuel, lubricating oil and water were on hand in copious quantities – it would run for days without attention. The majority of successful engines were simple in design and offered foolproof operation so that maintenance was within the scope of the farm worker, gardener or chauffeur.

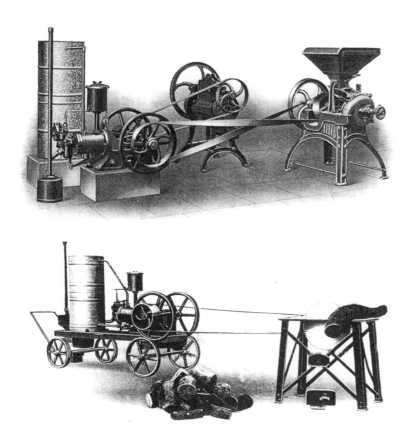

The stationary engine provided power from a fixed or 'stationary' position but if arranged on a truck or trolley, like this Petter 'Handy Man' oil engine, it could be easily transported to its place of work. The upper engine is driving a grinding mill and a cake-breaker; the lower one, which is portable, is driving a circular saw.

An engine for all purposes

A relative newcomer to the world of old stationary engines, if of an observant nature, might be wondering what the difference is between the humble stationary engine and the engines that powered the earliest motor vehicles (horseless carriages), tricycles, tractors and even motor launches. The answer, strange as it may seem, is very little. Although the similarity between all these modes of power is apparent before 1920, it is particularly marked the further back one travels towards the evolution of the fossil-fuelled internal combustion engine.

The stationary engine was the father of a curious mixture of offspring that would contribute towards the development of our society by propelling mankind on land, sea and in the air. Pioneers such as Otto, Day, Diesel, Akroyd-Stuart, Benz, Daimler and Priestman, along with countless others, all contributed towards building a vast industry based on the internal combustion engine. Each one of these innovators initially constructed a stationary engine in its own right.

The earliest prime movers, however unorthodox in configuration, were crude but functional, and easily adapted to sundry roles. With the simple addition of a few brackets, mountings and cables (for controls), the versatile stationary engine was converted for automotive or marine uses.

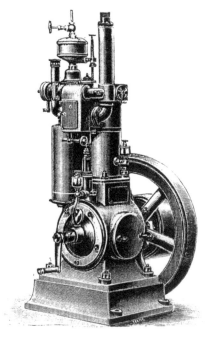

The Daimler of the early 1890s was advertised as a 'fixed engine' – the term 'stationary engine' came later. Not only was it intended to be used as a 'fixed motor' but also for driving fire engines, tramways, light carriages, quadricycles and even for launches.

In the pioneering period just before 1900 few firms were set up for the sole purpose of manufacturing gas and oil engines. Rather, the move into this field began as an offshoot from some other engineering line, perhaps agricultural, industrial or marine. For instance, an agricultural engineer could see the possibility of extra business by offering a suitable engine with his equipment – after all, other people were moving into this new market and this form of cheap power was catching on. Engines were somewhat of an unknown quantity at the time so a firm wishing to enter this market had two options from which to choose. The first was to look at the competition and produce something that looked along the same lines of the most successful, being careful not to fall foul of any patent litigation. Patent litigation was, needless to say, rife among engine manufacturers at the time. The second, and more expensive, method of entering the engine production market was to seek the advice and services of one of the engine inventors (or improvers), who might well be found to have a number of ideas that had been born and developed at a previous place of employment!

By the 1920s the majority of established engine manufacturers were working to their own designs. Owing to a rapid growth in road transport the scene was set to change with the advent of purpose-built engines. Fringe markets disappeared as manufacturers specialised, and suddenly there were marked differences between stationary, marine, automotive and aero engines. Unfortunately, the pioneer period had come to an end.

The trend, by that time (as it still is), was to aim towards lighter, faster stationary engines that would develop a higher

The earliest prime movers were crude but functional and so easily adapted for sundry roles – stationary to marine or automotive. Occasionally the role was reversed, as illustrated in this photograph of an early two-stroke marine engine, from the United States, which has been adapted for use as a stationary power supply.

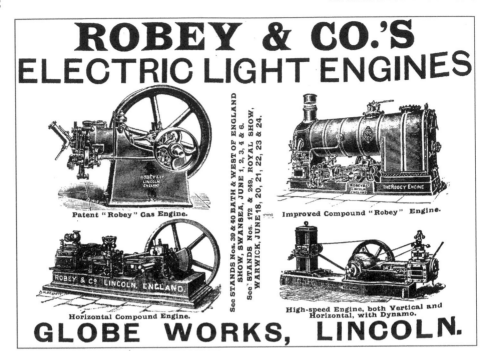

ROBEY & CO.'S
ELECTRIC LIGHT ENGINES

Patent "Robey" Gas Engine.

See STANDS Nos. 39 & 40 BATH & WEST OF ENGLAND SHOW, SWANSEA, JUNE 1, 2, 3, 4 & 6.
See STANDS Nos. 172 & 249, ROYAL SHOW, WARWICK, JUNE 18, 20, 21, 22, 23 & 24.

Improved Compound "Robey" Engine.

Horizontal Compound Engine.

High-speed Engine, both Vertical and Horizontal, with Dynamo.

GLOBE WORKS, LINCOLN.

Initially gas and oil engines were in direct competition with the steam engine but as the former became more efficient they usurped the latter with ease – internal combustion totally eclipsed external combustion. This 1890s advertisement shows a solitary Robey gas engine among its steam counterparts, but a decade later the steam engine would be in the minority.

level of efficiency and fuel economy. To attain this end, many manufacturers in the United Kingdom called on the services of Harry Ricardo (later to be knighted), who specialised in this field. His technique of creating combustion chamber designs that gave smooth and efficient running with both petrol and diesel engines – the latter being with indirect injection systems – won him worldwide acclaim in engineering circles. Interestingly, one of his better-known styles of combustion chamber has reappeared in Europe as motor manufacturers are turning towards the small diesel engine for use in mass-produced passenger cars. The success achieved by Ricardo's designs, inventions and modifications can be measured by noting the number of engine manufacturers that called upon his services. These include Lister, Tangye, Peter Brotherhood, Pelapone and Mirrlees Bickerton & Day.

Small-power engines

The newcomer, upon being confronted with an array of small-power engines for the first time, will probably regard them as working models. However, they were working engines in their own right. In answer to the question 'What did they actually do?', the stock answer is 'Anything within their power'. In terms of power, comparison can be made between the small-power engine and a fractional horsepower electric motor, both being able to perform the same tasks.

In the period when these engines were in almost everyday use, the amateur, or for that matter the professional engineer, would often use a small engine to drive his lathe. In all probability the lathe was treadled (foot-operated) in the first instance to machine a set of castings and parts to make a

The firm of Scott-Homer made quite a large range of gas engines, varying in size from ⅛ to 1 hp, supplied as castings or complete units. In 1910 a set of castings for a ¾ hp engine cost just over £2. Ignition was usually by hot-tube, the ignition tube being clearly visible in this illustration although the heating furnace is missing – it is normally attached to the rod that can be seen behind the tube. Its height is adjustable so that the actual point of heat on the tube could be varied to give some degree of advance and retard. The 'shaped' push rod was typical of this make, although it was not used on all the engines in the range. The Scott-Homer factory was based at Cradley, Staffordshire, where the firm specialised in the manufacture of tools, lathes, model-engineering equipment, small horizontal and vertical steam engines and small DC dynamos. Illustrated here is a ¼ hp Scott-Homer 'HM' type with a 1½ inch (38 mm) cylinder bore. Note the 50p coin as a size indicator.

The firm of Stuart Turner will be familiar to model engineers everywhere. As a young man its founder, Sidney Marmaduke Stuart Turner, designed his first model steam engine while looking after a large steam-generating plant at Shiplake Court near Henley-on-Thames. He then produced a set of castings and, after approaching 'Model Engineer Magazine' for a trade review, was inundated with orders. In the ensuing years the expanding business covered most requirements for the model engineer, ranging from steam and small petrol engines to electric motors and motor launches. The Stuart 800 gas engine seen here has a 1½ inch (38 mm) bore with a 2¾ inch (70 mm) stroke and is rated at ⅛ hp and listed as being suitable for driving a small lathe, dynamo or pump.

For the beginner Stuart Turner offered the 400 gas engine, sometimes referred to as the ⅛ hp gas engine. The castings were simple, with cylinder and body in one piece; all faces were prepared in order to obviate any complex machining. Electric or tube ignition was available, the latter being used on this particular example. The chimney, or furnace, can be clearly seen, its height being variable so that the actual point of heat on the tube could be adjusted to provide some degree of advance and retard. Note the 50p coin as a size indicator.

The same engine with the chimney or furnace removed to show the hot-tube. The height-adjusting rod, to vary the timing, can be seen more easily here.

In the 1920s Stuart Turner offered this 1/4 hp horizontal gas engine, naming it after its rated output – the 'quarter horsepower'. Available with tube or electric ignition, the cost was £5 in kit form. Bore size was initially a 1⁷/₈ by 2³/₄ inch (48 by 70 mm) stroke. The version illustrated here has the later 2 inch (51 mm) bore. Stuart Turner stated that it could be built entirely on a 3¹/₂ inch (89 mm) lathe. An almost identical 'big brother' came in the form of a 1/2 hp version, differing mainly in bore and stroke measurements.

suitable small-power engine to drive it in the future. A 1 hp engine could drive quite a large lathe, although engines with an output of only ¹/₄ hp drove many smaller ones. A suitable small-power engine could be built from a set of castings and parts frequently advertised (between 1895 and 1925) in engineering and hobby magazines. The rough set of castings enabled the buyer to machine each component on his lathe before assembly. Certain parts, such as piston rings, valves, springs, bolts and nuts, were supplied finished and ready to fit. Whether you bought these parts or not depended on your skills

The Tom Senior 'Simplex' was a very good example of a straightforward, easily understood design that worked well running on either petrol or gas. These engines were supplied as sets of castings or completely finished units and were produced in four sizes, ranging from 1/16 to 1/4 hp. Speed was hand-controlled and ignition supplied by trembler coil or magneto. A special cast-iron sub-base was produced as an extra to enable the magneto to be direct-coupled to the engine crankshaft, thus making a very compact power unit. The inlet valve used automatic operation, while the exhaust valve was push-rod operated. This type of engine was often incorporated as the main power unit in Senior lighting and charging plants.

Confusion soon occurs when trying to identify certain small-power engines because of a lack of maker's plates, serial numbers or even the odd logo. Although a number of makers offered their engines in complete built-up form, the vast majority of purchasers opted for the set of do-it-yourself castings, which did not include a nameplate – possibly for obvious reasons! This Tom Senior 'Superior' gas engine was built by the well-known engineer and model-maker H. Davey, who added his name, the date (1921) and the coat of arms of his home town (Lincoln) to the splash-guard.

Frank Hartop & Sons of Bedford made this style of engine during the 1920s and throughout the 1930s. Around five different types of engine were included in the range, two types being horizontal and three vertical, one of the latter being a two-stroke. This factory photograph depicts the 'S' type, which was available either as a set of castings or as a finished complete engine. A choice of ignition systems included magneto (as seen here), coil with wipe contact or hot-tube – termed 'porcelain tube'. Bore sizes ranged from 2 inches (51 mm) (1/3 hp) to 3 inches (76 mm) (1¼ hp).

as an engineer. When the engines were built up from castings, the more venturesome builders sometimes included ideas of their own to improve their engine's speed, efficiency or general appearance. While a member of the local engineering club might add a fancy side-shaft actuating an arrangement of rods and rockers backed up with a complex form of ignition system, the other extreme might involve someone with no expertise at all struggling to assemble an engine from the same set of castings. The result would be two engines bearing not the slightest resemblance in looks and performance. This explains why it is sometimes rather difficult to ascertain the actual maker of a small-power engine, particularly when working from a rather poor photograph. Though an engine may look like a certain make, one is left wondering why it has a different governing system or valve-operating arrangement from that specified by the maker.

It is important to remember that many of the firms that

This scale model of a National gas engine, with magneto ignition, was constructed to exhibit at model engineer exhibitions rather than perform an actual job of work. In the engine heyday apprentices learning their trade with the larger manufacturers often used their spare time to build a scale model of one of the current types of engine in production at the time. Some of the models were so accurate that without the use of a size indicator (a screwdriver in this instance) it would be impossible when looking at a photograph to distinguish them from the real thing.

advertised these sets of castings were simply mail-order engineering and model shops. The name of the kit was either the distributor's own name or a brand name. For this reason engines that have been built from one maker's castings often appeared under an array of names, thus causing confusion for collectors and enthusiasts.

Small-power engines usually used ignition systems that were scaled-down versions of those used on full-sized engines in use at the time – the only exception being the magneto, as it was not a practical proposition to manufacture this part in miniature. Some arrangement of trembler coil was most favoured. A variety of suitably sized carburettors was usually available, ranging from plain needle-valve types to some quite elaborate float-feed styles. Many of these engines were arranged to run on the town gas used for lighting purposes and in this respect were equipped with some form of extra admission valve and a mixer-type carburettor. Governing systems, when fitted, were usually smaller copies of the various types that were fitted to full-sized engines.

There were several well-known makers of small-power engines, the most prominent of the day being Stuart Turner Limited, Hobbs Motor Company Limited, Scott-Homer of Cradley and Frank Hartop & Sons. Although the engines of these manufacturers are, in most cases, instantly recognisable there are certain other makes of which no two examples appear to be the same. Typical makers of such wildly varying products include Henry Butler and Madison Castings of Derby, both firms being interconnected around 1900. A further example that is conspicuous in lacking uniformity of any kind is affectionately termed 'Leek' by enthusiasts after the maker British Engineering & Electrical Company of Leek, Staffordshire.

Major manufacturers

Amanco

In a poll run by *Stationary Engine Magazine* the most collected make of old engine was found to be the Lister – presumably because of the considerable number of these engines available. If you find an old engine you can quite safely assume that it will probably be a Lister. For similar reasons the Petter came second, while in third place was the Amanco. The Associated or, as they are commonly known in the United Kingdom, 'Amanco' engines were made by the Associated Manufacturers' Company, Iowa, United States. Without doubt the indomitable

Ripe for restoration, just how enthusiasts like to find them! The author purchased this Amanco 'Hired Man' in the 1960s for £5 – a considerable amount at the time. A petrol/paraffin (kerosene) version of around 1923 vintage, it has the 'goose-neck' picker bracket and 25R magneto and is fitted with a pulley to drive sheep-shearing equipment. From 1916 until well into the 1920s these engines were known, in the United States, as the Iowa 'Oversize' because under test conditions at the factory they developed some 25 per cent over their rated horsepower.

This restored Amanco 'Hired Man' is an early 2¼ hp version with rolled, split water hopper with rectangular filler. Another early feature is the bell-mouth mixer (carburettor) and straight-arm igniter picker. It is either one of the first to use a skew-gear-driven magneto (instead of taking its drive directly from the crankshaft via intermediate gearing) or it has been modified during its working life. An unusual feature of early Associated Manufacturers' engines is the way certain components were painted silver in order to dissipate heat and assist cooling.

This unrestored 'Three Mule Team' Amanco was supplied second-hand in the late 1920s and was last used in the 1970s to drive a saw-bench. This example is unusual in having a front-mounted fuel tank, as the 3 hp tends to utilise a rear-mounted tank in the manner of the larger engines in the range. The 'Three Mule Team' name was quickly phased out in the United Kingdom in favour of simply 'Amanco'.

Amanco is one of the great success stories in the world of stationary-engine marketing as thousands were imported into the United Kingdom. In general these engines were crude, obsolete in design, with no concessions to anything but utility. Their finish was, in general, abysmal in comparison to that of British-built engines. So why did the Amanco sell so well? There are several reasons for this: it worked and kept on working; the important parts lasted; the material, although slung together, was well machined, jigged and filled; and, most important of all, it was cheap. The Amanco was far more affordable than the Blackstone, the Crossley or the Ruston, with their coats of gleaming enamel so carefully applied on castings with blow-holes painstakingly stopped and filled and with a lot of costly and unnecessary machining on non-functioning surfaces. The Amanco was cheap because its mechanism was the simplest possible, with no complex valve gear, rotating side-shafts, complicated lubrication system, expensive brass lubricators – in fact nothing but the basic requirements. How could these engines fail when they arrived in Britain adorned with catchy names promising such a labour-intensive service – 'Chore Boy', 'Hired Man', 'Hired Hand', 'Farm Hand', 'Six Mule Team' and 'Foreman'? The descriptions were admirable; the engines performed as promised and, furthermore, the price was right.

The large Amanco engine embodies a threatening presence that has not diminished with the passing of time. This one has the early igniter striker arm and original cross-drive magneto that identify it as an early example. The 'Foreman' (8 hp) and the 'Twelve Mule Team' (12 hp) were among the larger of the Associated Manufacturers' Company engines. These are quite rare in the United Kingdom. In 1915 a colossal 'Eighteen Mule Team' was advertised, topping the range in terms of power.

In the late 1920s the Bamford OV (affectionately known at the factory as 'old vertical') petrol engine was introduced in 3, 4 and 5 hp sizes. An unusual feature is the circular fuel tank mounted on top of the water tank, although a crescent-shaped tank mounted against the tank itself superseded this. Wico EK magneto ignition is used.

Bamfords

The company Bamfords Limited was supplying agricultural machinery as far back as the 1870s. The decision to manufacture a line of stationary engines to drive their extensive range of barn machinery was taken immediately after the First World War. The first engines were of horizontal design: a 5–6 hp paraffin model, followed by a 2½ hp petrol engine, appeared in 1920 and immediately gained a silver medal at the Darlington Royal Show. The arrival of these engines provided a new direction for business in the form of lighting plant production. The first of these was a small 2 kW version and then a range of saw-benches came on to the market. Following the general trend set by other manufacturers, Bamfords Limited moved into the diesel engine market in the 1930s with 6, 8 and 10 hp models using indirect injection – the combustion chamber design coming (under licence) from Russell Newbery. Specialities linked with Bamfords' engines are caged valves and split-hinged crankcases, both intended to reduce working costs, servicing and overhauls. Engine manufacture continued until well into the 1970s but the company went into voluntary liquidation in 1980.

The Bamford EV1 was in production from 1929 until 1946, a versatile engine designed for the agricultural market – its main rival being the Lister 'D' type. A useful feature was the hinged crankcase allowing for easy inspection and maintenance. This well-used, tired, but mostly complete example is how enthusiasts like to find an engine.

After being superbly restored, the same Bamford EV1 is displayed with a Godwin water pump. The engine does line up with the pump – it is cleverly mounted on a swivel base to facilitate storage and transport. Two styles of eye-catching transfer were used by Bamford – a 'silver medal' version (seen here) commemorating the medal awarded to the company at the 1920 Darlington Royal Show, and the 'seven colour' type.

Bentall

E. H. Bentall & Company of Maldon, Essex, had a colourful history in agricultural engineering that dated back to the 1790s but, rather like Petter (see page 28), it was a failed attempt to diversify into motor-vehicle production that suggested stationary-engine manufacture as a viable alternative. Unlike Petter, which produced only a handful of cars, Bentall did at least sell around one hundred between 1906 and 1913 – but it is the company's stationary engines that are of interest here! A range of heavy-duty vertical engines appeared in 1909–10, followed by a series of horizontal open-crank engines in 1912. Engine production continued through the 1920s but by the 1930s it had ceased and the engine patents were sold to Petter. Manufacture of the well-known Bentall engine valves began as far back as 1904 and represented an important product of the Heybridge factory for many years.

Right: The robust Bentall vertical was available in 2½, 3, 4, 5, 7 and 9 hp sizes in petrol and paraffin forms. It was unusual for a British engine to have overhead valves at that time (1909–10) and, furthermore, it was among the first to have an impulse arrangement for use with an HT magneto in order to obtain a higher armature speed and thus give a better spark for starting – this device appeared as early as 1912. This 5 hp petrol engine was supplied new in 1912.

Left: Also from Bentall came the 'Pioneer', an advanced (for the time) hopper-cooled horizontal with HT ignition and mechanically operated valves. Unfortunately the external oil reservoir (mounted on the water hopper), designed to feed the various moving parts, was the engine's Achilles' heel. Lack of oil replenishment and servicing led to premature demise. Any form of sophistication was anathema to the farming community, which tended to opt for cheap imports from the United States such as the Amanco, sold in Britain by the thousand.

Blackstone

Another company with a history steeped in agricultural engineering to enter the stationary-engine market was Blackstone. The firm came about in 1895 when Mr E. C. Blackstone took notice of a newly marketed paraffin engine being made by four brothers in Billingshurst, Sussex. Following an agreement made the following year, the 'Reliance' engine, as it was called at the time, and its two inventors, Frank and Evershed Carter, arrived at Stamford, Lincolnshire, to develop a Blackstone version of their own design. The first Blackstone engines were of a type that required a constant flame in order to keep the vaporiser and ignition tube hot enough regardless of whether or not the engine was running at full power. Within a short period, following the general trend of the time, engines with automatic ignition appeared, needing only a blowlamp to get them started.

The majority of successful engines tended to evolve over a period of time rather than suddenly appearing on the market. The Blackstone is a prime example of this, with modifications being added to normal production engines so that each type or style to appear was an updated version of the original. This is exactly what happened, between 1903 and 1905, to the Blackstone engine that became known as the 'standard model' and that formed the basis for a very successful range of engines that remained in production for over twenty years.

In 1923 a 'spring injection' oil engine superseded the older models, followed by a range of petrol/paraffin engines. As was the general trend in engine production at the time, a small vertical diesel engine was introduced in 1932, followed in 1935 by the 'DB' (commonly dubbed the 'diesel box'). The company became involved with the ill-fated group Agricultural and General Engineers after the First World War and, in 1936, one hundred years after its formation, the company was bought by Lister, an arrangement that spawned the trading name Lister-Blackstone. In later years, along with Lister, Petter and others, the company became part of the Hawker Siddeley conglomerate.

The first of two distinguishing features of the Blackstone 'standard model' was the Carter's patent governor that appeared in 1903. The other was the horizontal vaporiser of the closed type fitted with an internal igniter – differing from most other makes of oil engine in having ignition controlled by a timing valve. This was used to regulate the period of ignition by cutting off the heated parts during compression – hence obviating pre-ignition, which tended to damage hot-bulb engines. The standard model remained the Stamford firm's basic oil-engine design for over twenty years.

The Crossley 1030/1040 and 1050 range was heavy, compact, of simple design and with no concession to appearance. Available in petrol/paraffin and gas format, this engine came out late in 1925. Both valves are tappet-operated from a single rocker arm, which is carried on a bearing mounted on the cylinder head. The magneto is conveniently mounted at the end of the crankcase on a bracket. The range consisted of three sizes – 2/2½ hp, 3/3½ hp and 6 hp, with an operating speed of 700 rpm for the smaller sizes and 550 rpm for the larger.

Crossley

Crossley Brothers Limited can be listed as one of the pioneering engine producers in Britain. The company was founded in 1866 by brothers Francis and William Crossley, who, by either a stroke of luck or extreme foresight, acquired the British patents (in 1869) to build and develop the novel Otto & Langen gas engine. Seven years later, when the Otto Silent gas engine arrived, using the new four-stroke cycle, Crossley Brothers Limited immediately commenced manufacture of its own version, the Crossley 'Otto' gas engine. A slide valve controlled the admission of the gas and air mixture into

Right: *The double side-shaft Crossley PH (stationary) or PP (portable) 1060-1075 was intended for use in the farming environment but proved to be over-engineered and consequently too expensive for the average farmer. It was available in petrol, paraffin or gas format, and output extended from 3¼ to 7½ hp. This is a nice engine for the enthusiast to own, being well finished with plenty of moving parts. An air silencer provides quiet running too.*

The Crossley 1½ hp type 'H' gas engine with tube ignition. Although it was only partly restored when this picture was taken, already some three hundred hours had been spent in fabricating missing parts – such is the dedication of the old engine enthusiast.

Finding an old stationary engine much bigger than this gigantic 'girder frame' Crossley is somewhat unlikely nowadays. Built around 1910, it belongs to what its maker termed the 'Z' series and gave an output ranging from 107 hp on suction (producer) gas to 125 hp on town gas. Bore and stroke of 19 by 27 inches (48.2 by 68.5 cm) plus 84 inch (213.2 cm) flywheels give an idea of its size. The engine spent its working life driving joinery plant in Australia and although it looks somewhat neglected it still runs.

the cylinder while employing a system of flame ignition. This mode of ignition became known as the 'carrier flame' type. In 1888 poppet valves were introduced for gas and air admission and other improvements quickly followed, including tube ignition and, later, electric ignition. More than 100,000 Crossley gas and oil engines were built, many examples of which still remain. An extensive range of petrol, paraffin, two-stroke and diesel engines followed, catering for all aspects of agriculture, industry, automotive and marine markets.

Lister

The vast majority of newcomers to engine collecting in the United Kingdom start with a Lister as their first engine, usually a model 'D'. Few manufacturers can claim to have produced an engine that remained in production for as long as forty years. R. A. Lister & Company Limited can, however, justify such a claim as the firm's successful general-purpose 'D' type, introduced in 1926, was still being produced, albeit in small numbers, as late

Right: The Lister 'D' type has been a familiar sight around farms, contractors' yards and engine rallies for many years. Furthermore, a large number of enthusiasts (the author included) were introduced to the old-engine hobby after finding and restoring one of the 250,000 produced. This versatile engine occupied diverse roles, ranging from generating electricity (seen here) to pumping water and driving milking machinery.

Right: While the vast majority of Lister 'D' types are hopper-cooled, purchasers could specify tank or radiator cooling, with the latter (shown here) using a belt-driven fan.

Left: *The 'D' became the 'DK' when equipped to burn paraffin or (in later engines) TVO (tractor vaporising oil). Original factory-supplied 'DKs' used a flat-top piston to provide the acceptable compression ratio.*

The 2¹/₂ hp and upward 'A' and 'B' type is one of the better-known and more easily recognised Lister engines. Unlike the 'D' type, which had overhead valves, the 'A' and 'B' had valves mounted in the side position – initially in easily detachable cages positioned under the cylinder head. Later the valves were moved to inside the water hopper under a pair of brass plugs, which are often difficult to remove. However, in common with the 'D' type, there are many confusing variations – few Lister engines are identical in every respect.

Left: *Lister petrol engines, in single-cylinder form, do not come much bigger than the 9 hp 'L' type known among enthusiasts as the 'L31'. These engines were supplied with a vast 6 gallon (22.7 litre) capacity water hopper instead of the conventional tank cooling in order to save valuable installation space in machines such as stone crushers, cranes and concrete mixers. This 1936 example was built towards the end of a lengthy and successful production run that was curtailed by the arrival of a more fuel-efficient range of diesel engines.*

Right: *The 'Q' type Lister is the paraffin-burning version of the 'R' type petrol engine, the largest in the range. Although most were rated in the 7¹/₂ to 8¹/₂ hp range, a few high-output 10 hp versions were built to special order. The determining feature, the vaporiser, is clearly visible on this engine, as is the screen-cooling arrangement.*

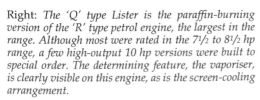

as 1964. Although a few major, and many minor, changes were made to this noble little engine over the years, it retained the same basic design throughout its production run of over a quarter of a million units. But it did not all begin with the famous 'D' type. Enthusiasts of the Dursley product will undoubtedly associate 1908 and 1909 with the development and introduction of the first Lister stationary engine – an event worth the highest possible recognition because the stationary-engine movement would not be the same today without its thousands of Lister engines, an example (or two) of each residing in most engine collections!

The next best-known Lister is the 'A' type, which was introduced in 1923 – another mass-produced example surpassed in number only by the 'D' type. The 'B' type can be looked upon as the slightly bigger brother of the 'A' type, as the difference between the two models at the time of inception was just 1 hp. In terms of production numbers the 'A' type far exceeded the 'B' type, but this is not an adequate explanation for the lack of 'B' types seen at stationary-engine rallies.

Of the 'H' to 'R' types – a range dating back to the first production in 1909 – the 'H', 'J', 'L', 'N' and 'R' are the petrol variants while the 'K', 'M', 'P' and 'Q' were oil burners. Some engines were arranged to burn zero-octane low-grade illuminating oil called 'paraffin' or 'kerosene'. These engines were fitted with some form of exhaust-heated vaporiser and generally required a lower compression, obtained by a flat-top piston. A slightly faster running speed was another feature – to aid combustion and to recover a little of the power lost by lowering the compression. During the late 1940s and 1950s TVO (tractor vaporising oil) was widely used, this fuel being slightly more volatile, with a 50 octane rating. Various

For the Lister enthusiast wanting something different, the 'G' type is worth considering for the number of variations likely to be encountered. Termed an 'industrial stationary engine', the 'G' type tended to be built in batches to a specification dictated by the customer – hence we come across many special components such as air filters, cowlings, clutches, gearboxes, screened ignition sets, tropical radiators, extended hand controls, extended or shortened crankshafts and extra flywheels. Many were supplied to the Ministry of Defence; again these were to a variety of specifications. This 'G' type was supplied new in 1940.

Not all Listers are of Dursley origin. These air-cooled Wisconsin engines were initially imported from the United States either as complete units or in parts to be assembled at Dursley. Later they were produced almost entirely at the Lister factory. This Lister AFL 50 has 'Wisconsin' stamped on its piston skirt. It is one of a batch built in 1940.

manufacturers were quick to provide cheap TVO conversion during the post-war period of the late 1940s when petrol was in short supply. A conversion kit comprised a vaporiser, a container in which to carry a little petrol for starting purposes, a three-way tap and an extra cylinder-head gasket to lower the compression.

Petter

It was a series of articles entitled 'How to make a small gas engine' appearing in the 1895 *Boy's Own Magazine* that gave Percival and

As found! This Petter 'Handy Man' type HF 2¼ hp (test-bed rated at 3 hp) was supplied new to agricultural agent Berkley Uncles of Bradford-on-Avon, Wiltshire, in October 1906. The 'Handy Man', designed in 1902 and introduced in 1903, is a cheaper version of the Petter Standard Series, the latter proving too expensive for the farming community, which preferred less expensive American engines. This style of Petter was a development of the 1896 horseless-carriage engine. Starting is by blowlamp heating an ignition tube that projects from the vaporiser or hot-bulb. Once the engine is running and sufficiently warm, the blowlamp can be extinguished owing to the operation of a secondary tube (called an internal tube), which is situated in the lower region of the vaporiser. This internal tube remains heated by hot exhaust gases being directed upon it and thereafter retains enough heat to ignite the vaporised oil, aided by heat generated by compression. Fuel (lamp oil or paraffin) is carried in the pedestal-mounted cylindrical cast tank and fed by gravity to an adjustable drip-valve. A belt-driven governor takes its drive from the crankshaft and is so arranged as to supply greater or lesser quantities of oil and air as required.

The Petter 'S' type surface-ignition oil engine is a close descendant, and very similar in design, to the earlier range of two-stroke engines to come from the Yeovil factory. The 'S' type designation was adopted in 1923 when the Petter patent cold starter was introduced. This device consisted of a combustible cartridge that was screwed into a holder in the vaporiser. Once lit, it generated enough heat to ignite the preliminary charges. The 'S' type was available in single- or twin-cylinder form, the singles being produced in 5–6, 8–10, 12–15 and 18–21 hp. Production ceased around 1940. Here we see the smaller 5–6 hp and the larger 18–21 hp version.

Ernest Petter the idea of making an engine that could be used to drive a horseless carriage. Although the 3 hp hot-bulb engine was a success in its own right, the task of preparing it for automotive use proved insurmountable because of a lack of funds and expertise. In an attempt to extricate themselves from what might have been financial ruin, the Petter brothers changed direction completely and adapted their oil engine for agricultural and industrial use. This brilliant decision culminated in the launch of a succession of hot-bulb, four-stroke stationary engines, which remained very similar in style until their demise in 1915.

In 1909 another vital decision was made, this time to introduce a range of vertical two-stroke engines while retaining the hot-bulb (though using slightly higher compression) form of ignition. These engines were initially described as 'semi-diesel' but later as 'surface ignition' in an attempt to phase out the 'diesel' name because of its German connotations.

A magneto (spark) ignition petrol/paraffin engine, designed as a cheaper alternative for the farming community, appeared in 1913. Initially called the 'Petter Junior' but later dubbed the

Left: *The following terms can all be correctly used to describe the smaller of the Petter family of two-stroke engines: 'Petter Junior', 'Little Pet', 'Petter Universal', 'M' type, 'VZ' and even, in some guises, 'Handyman'. The name 'Little Pet' was phased out after someone explained to Percy Petter (at a Paris show) that 'pet' means 'fart' in French! Introduced in 1922, the small 'M' type remained in production until 1939. Before 1931 the main bearings were lubricated from oil wells from which the oil was carried to the shaft by a rotating ring. Later engines used roller-bearing mains. First-series engines, easily identifiable by the front-mounted magneto, tended to be troublesome and were soon superseded.*

Above: *A popular use for Petter two-stroke engines was supplying electricity for home, farm, shop, factory or more mundane purposes such as automotive battery charging in a village garage. In the 1930s Petter offered a profitable business opportunity by providing one of these sets especially set up for charging radio batteries – in that archaic period before dry batteries were readily available. Petter-Light plants were available either with belt drive or with a dynamo direct-coupled to the engine shaft – as seen here. Cooling was by hopper, water tank or radiator; sometimes a combination of two methods was used.*

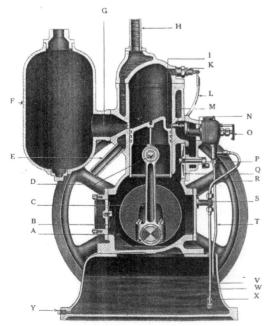

Left: *Sectional illustration of the Petter 'M' type engine. Key: A, crankcase; B, air inlet; C, connecting rod; D, piston pin set screw; E, water inlet pipe; F, exhaust silencer; G, exhaust ports; H, water outlet pipe; I, inlet ports; K, sparking plug; L, sparking plug terminal; M, fuel nozzle; N, fuel chamber; O, fuel regulating needle; P, air baffle lever; Q, air baffle; R, fuel delivery pipe; S, fuel pump T piece; T, fuel drain pipe; V, fuel; W, oil suction pipe; X, foot valve; Y, bed drain plug.*

Few other makes of engine were adorned with as many styles of water hopper as the Petter 'M' type; the terminology used by enthusiasts to describe these engines usually emanates from the shape of hopper used. The spherical hopper on this 5 hp Petter two-stroke is usually referred to as an 'apple top'. Engines made before 1923 used Stauffer greasers to lubricate the main bearings. This system was superseded by one using oil baths to lubricate rotating rings, which in turn was replaced, in 1931, by one using roller bearings.

'Universal M' type, this range of two-stroke engines was produced until around 1940. Interestingly, enthusiasts tend to identify this range by the shape of the water hopper, so we get: pear, ball, spherical, rugger ball, round, bucket, flat, oval, parallel-sided, prune, basin, jerry pot, wedge, skip, lavatory cistern, and so on – the jargon of a hobby that conjures up a pleasant way of life for the dedicated but means nothing at all to the layman!

Left: *The Petter Universal PU8, a four-stroke flat twin produced between 1933 and 1947, was sold in large numbers through government contracts and earned a reputation for driving boats in the D-Day landings and for crossing the Rhine during the liberation of Germany in the Second World War. Ministry of Defence generating and pumping sets were supplied in a carrying cradle (seen here) or a heavy-duty frame. A standard feature of ex-military sets is a pair of fitted boxes, one for tools and the other for spares, intended to keep the electrics in working order. Although the PU series went out of production in 1947, many were retained by the Ministry of Defence to power fuel pumps and generating plants and were brought back into service during the Falklands War. The PU8 seen here is the 8 hp 1800 rpm version, in the military cradle and with a flexible shaft drive, supplied to the Chief Inspector Engineer Signal Stores in 1940.*

Many examples of the mass-produced Petter 'A' series, an air-cooled four-stroke, can still be found either in use or at least in serviceable condition for just a few pounds. The first series, introduced in 1936, was rated from 1½ to 3 hp and was available in petrol and petrol/paraffin styles. It was so successful that a second series, introduced in the mid 1940s, extended to an output of up to 10 hp. A third and final series arrived in 1951. This factory photograph shows an 'A' series Petter in a typical 1950s application – arranged to drive a portable air-compressor set for use in a garage.

A superb example of an 1890s Hornsby-Akroyd. This style of engine was able to dispense with flame ignition, hot-tube or any form of electric ignition in favour of a charge of oil that was injected directly into compressed air in a red-hot vaporiser. The charge was fired automatically.

Ruston & Hornsby

When Richard Hornsby & Sons Limited produced an oil engine, under the famous patents (described earlier) of Herbert Akroyd Stuart in 1891, the neighbouring firm of Ruston Proctor & Company Limited was also in the process of developing an oil engine. Both concerns continued their independent engine developments until they amalgamated in 1918, becoming Ruston & Hornsby Limited.

Akroyd Stuart's engine, known as the 'Hornsby-Akroyd', was the first marketable airless-injection hot-bulb prime mover to achieve success. Other countries, such as Germany, Sweden, Norway, Denmark and Russia, soon copied the principle of injecting the fuel into the hot-bulb vaporiser. Gas-engine manufacture was undertaken from 1905 following an amalgamation of Hornsby and J. E. H. Andrews & Company of Stockport.

In the ensuing years the company serviced the whole spectrum of the market with a variety of engines. Particularly interesting were its large multi-cylinder vertical engines of the hot-bulb type operating on the four-stroke cycle in the early 1900s. This method was in stark contrast to the two-stroke

The Grantham-produced (in Hornsby's old factory) Ruston & Hornsby 'M' type was a further, and later, development of the original Akroyd engine. This 25 hp version was installed in a soap and polish manufacturing factory in 1925. It was started by compressed air and used magneto and petrol until thoroughly warm, when a change-over to fuel oil was made. The magneto and lead can be seen to the right of the governor.

The Ruston & Hornsby class 'PR' ('push rod') was introduced in 1922 and made in four sizes ranging from 1½ to 6 hp. Three styles of magneto were used during the production run. An air-cooled version was also available.

method employed by other manufacturers for hot-bulb ignition. It must be said that the four-stroke hot-bulb multi-cylinder vertical was probably not a great success. From the 1920s Ruston & Hornsby multi-cylinder vertical engines changed with the times by adopting compression ignition. Although an equally successful range of large horizontals augmented the verticals, it is the collection of smaller petrol/paraffin engines that will probably be of greatest interest to an enthusiast.

Above right: *Before restoration! A Ruston & Hornsby class 'PT', produced between 1936 and 1950. The first engines used a flywheel magneto but during the Second World War a change was made to the Lucas RS1, and later (as seen here) came the Wico 'A' series magneto. Power ranged from 1½ to 4 hp (700 to 1500 rpm). Interestingly, when 'PT' production ceased Ruston & Hornsby passed the design to Wolseley (1951) and soon after a PT look-alike appeared in the form of the WD8.*

Right: *After restoration! Although green was the general colour for the Ruston & Hornsby class 'PT', this 1947 version is one of a batch to be painted Hornsby maroon – reputedly because of a post-war paint shortage. At this time, and for similar reasons, the fuel-tank cap was changed from brass to plastic. Early engines used Ruston's own carburettor, later to be superseded by an Amal type.*

Wolseley

The name of Frederick York Wolseley is synonymous with sheep-shearing equipment – the stationary engines came later! Wolseley's first ventures into sheep-shearing equipment took place in the 1860s while he was managing a shearing station in Victoria, Australia. Numerous experiments took place but Wolseley was constantly exasperated by the lack of local engineers capable of putting his ideas into practice. For this reason he was forced to return to England in the early 1870s to seek assistance, although he was back in Australia again by 1874. In 1887 the Wolseley Sheep Shearing Company was founded and it was about this time that Herbert Austin (later to be associated with the car of the same name) joined the new shearing business. Around 1890 Wolseley transferred the business to England, initially to London but in 1893 to

A typical sheep-shearing outfit as marketed and exported by Wolseley from around 1909. It was portable, compact and ideal for the sheep stations of Australia and New Zealand.

Right: *The Wolseley WD1 was built following a request by the War Department (hence 'WD') for a batch of small engines for use in the desert. Engines were placed on the home market in 1943 and continued in production until 1945, by which time 3548 had been built. The WD1 was replaced by the finned-hopper WD2, which was deliberately aimed at the farming community. Being built in a post-war environment helped to increase production substantially to over seventeen thousand units before this model also was superseded by a later version – this time the WD8.*

Birmingham – the workshop of the world. In 1901 the company sold its up-and-coming motor business to Vickers Sons & Maxim; later the car business went to Morris. Herbert Austin at first remained a director of both Wolseley and Vickers but subsequently left to form his own car company.

From these chequered beginnings it is apparent how Wolseley came to have a ready-made market for sheep-shearing equipment and cream separators in Australia. As soon as the first stationary engines were built at the new Alma Street works they were exported along with shearing equipment. Although engines were supplied with lighting plant, pumps and general farm appliances, this was a minority market for the company and remained so until the late 1920s.

Above left: *Before restoration! Sometimes a water hopper did not have enough capacity to cope with an engine running unattended for long periods. This Wolseley WLB1 of 3½ hp had been modified, during its working life, to provide extra capacity by means of a plate enclosing the hopper. A further plate screwed over the filler aperture carried a pipe extension for tank cooling.*

Above right: *After an extensive rebuild! The Wolseley WLB1 was built between 1946 and 1949 with a meagre 691 examples leaving the factory.*

Lesser-known manufacturers

Left: *A popular engine among enthusiasts is the Victoria, made by the Bristol Wagon & Carriage Works Company Limited, Bristol. Built in a variety of sizes, in petrol or paraffin format, production began in 1906/7 and continued until the early 1920s. The 5 hp model seen here is being refuelled by Eric Brain, who has carried out extensive research into the background behind the Victoria engine and its Bristol-based manufacturer.*

Above: *The robust Powell is a typical example of a heavy-duty farm engine. Powell Brothers Limited, Wrexham, entered the engine market at a particularly difficult time following the First World War and continued until liquidation curtailed production in the mid 1920s. Cudworth & Johnson Limited, also of Wrexham, then took up Powell engine production, albeit for a short time. The Webster magneto and 'arm breaker' starting handle (secreted in the flywheel rim) are typical Powell features.*

Of the various makes of air-cooled engine in preservation, the Coborn offers a little rarity. Introduced in the early 1930s by Kryn & Lahy Limited, these engines were later manufactured by Browett Lindley Limited and later still by K. & L. Steelfounders & Engineers Limited – all being members of the '600 Group' of companies.

Above: *While certain makes of engine remained unchanged in basic design for years, the diversity of others creates a nightmare for the enthusiasts who own them. A typical example of one such engine is the 1922 design Hobbs, built in Frome, Somerset, until 1928, when the engine business was transferred to Crawley Agrimotor Company. This example of a 1922 design Hobbs was probably one of the last engines to be built in Frome before the move. In the 1930s Agrimotor sold these engines with a range of Teles drag-saws.*

Above: *Tangyes Limited's first engines (from around 1881) were mainly experimental products developed by contemporary pioneers such as James Robson and Dugald Clerk. But with the arrival of the four-stroke mode of operation, Tangye, in conjunction with C. W. Pinkney (a design engineer), produced a gas engine, followed in 1892 by an oil engine. This heralded a successful line of engines, continuing until the 1960s. The example seen here is a Tangye lamp-start oil engine.*

Fowler of Leeds produced a versatile range of engines designated the 'P' type ('P' for 'petrol' and 'PP' for 'petrol/paraffin'). The four basic styles were divided into no fewer than fifteen sizes – each providing a different output to cover most applications. While the smaller engines in the range can obviously be classed as general purpose, the larger engines were heavy and intended for industrial use. Fowler claimed to have supplied batches of every size for use by the armed forces during the Second World War. This factory illustration shows a radiator-cooled '1PAM' supplied to British Asphalt & Bitumen Limited in 1938.

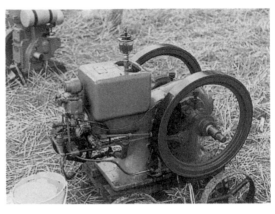

Right: *A popular import from the United States was the International Harvester 'M', with the first engines arriving in the United Kingdom in 1919. Although well received, the 'M' type did not usurp the well-established Amanco, yet in its country of origin it was a top seller with over four hundred thousand examples being produced between 1917 and 1937. Company policy dictated the continued use of well-known names swallowed up by merger and acquisition, hence we often see McCormick-Deering engines badged also with the name of IHC (International Harvester Company) – with one name on one side of the water hopper and the other on the opposite side.*

Left: *Ask an enthusiast to name any make of French engine and the answer will almost certainly be Bernard, or possibly Conord – a Bernard brand name. Bernard Engines & Company Limited was formed in 1919 and the following year the type 'D1' (seen here) made its appearance. Bernard engines were sold in the United Kingdom in the 1930s by Southwell of Worthing and by A. C. Bamlett, the latter company installing them to drive various items of farm machinery. In the post-war years Tarpen Engineering Company marketed various types of Bernard with pumping and generating plant. The newcomer will soon notice how the French have a definite predilection for copper and brass external fittings, with radiators being works of art.*

Right: *The well-known agricultural implement makers Ransomes, Sims & Jeffries Limited of Ipswich introduced the 'Wizard' in 1921. Here we have a rare start-from-cold paraffin engine (with no form of electric ignition) that uses a perforated pre-combustion chamber in order to fire the initial charge of fuel – a similar method to the air blast of the early diesel engines. The Bronz engine, built in Switzerland, was one of the earliest engines to use this method of combustion/starting, which was developed extensively and successfully by Daimler-Benz. When this photograph was taken this particular 'Wizard' was owned by enthusiast and old-engine photographer Philip Thornton-Evison, who is seen making running adjustments.*

Emerson-Brantingham is probably better known for its range of tractors than for stationary engines. The company, based at Rockford, Illinois, United States, was taken over by J. I. Case Threshing Machine Company in 1928. 'E-B' engines are not common in Britain and not plentiful in their country of origin either. The model 'U' seen here, made between 1914 and 1920, is a typical example of a crudely built American farm engine. Ignition is provided by the front-mounted Webster tripolar oscillator.

The 3 hp air-cooled Glasgow was introduced in 1920 by Wallace (Glasgow) Limited, maker of the Glasgow tractor. It uses the Burt-McCollum single-sleeve arrangement, whereby the sleeve is operated by a short connecting rod driven from a small crank from the half-speed shaft. The single sleeve-valve operation combines rotary and reciprocating motions and is consequently never at rest – a feature claimed greatly to assist lubrication. Engine production was short-lived, terminating in the mid 1920s, although advertisements clearing old stock ran until the end of the decade.

Above: *Few firms set up business with the sole intention of building a stationary engine; to most it was either an offshoot of an already existing line of business or, more often, a stopgap in times of recession. Hewlett & Blondeau, an aircraft manufacturer based in Leagrave, Bedfordshire, came into the market-place after the latter turn of events. Having set up in 1914 to build Farman aircraft, the company initially flourished, thanks to government contracts, during the First World War. But upon cessation of the hostilities the firm found itself on the brink of recession, with a large workforce to employ and 12,000 square feet (1115 square metres) of building to utilise. A desperate attempt to change direction by switching manufacture to farm engines produced the 'Omnia', with the intention of putting some five thousand engines on to the market. The first engines arrived in 1919 but by late 1920 the firm had succumbed to post-war recession. Companies associated with aero-engines have a reputation for over-engineering stationary engines (Armstrong Siddeley being a prime example) so it is somewhat of a surprise to see the 'Omnia' introduced with an outdated low-tension ignition system – although a change was quickly made to incorporate a high-tension system (with sparking plug), as seen here. These throttle-governed petrol/paraffin engines are exceedingly rare as few seem to have survived.*

Morton engines, named after the Yorkshire village in which they were made in small numbers, are also exceedingly rare. This large oil engine was probably converted to high-tension ignition at some stage during its working life.

Nayler & Company Limited of Hereford was founded around 1880 as a general manufacturing engineer. The business offered an extensive range of products from machine tools to steam tractors, wagons and pumps. Two-stroke vertical stationary engines were offered as well as a variety of horizontal lamp-start oil engines, the latter with such flamboyant names as 'Bullock', 'Bugler' and 'Culprit', or 'Buglehorn' for the extremely rare example illustrated here.

Left: A surprising number of motorcycle or motorcycle engine manufacturers dipped a foot into the stationary-engine market. While some, such as JAP, Villiers and BSA, were successful, the vast majority were small-time producers. A typical example from the latter category is Vincent, a well-known Stevenage-based company renowned for its world-class motorcycles. Several attempts to enter the stationary- and marine-engine market were beset by financial problems. The various Vincent stationary engines, although well built, lacked research and development as well as any firm marketing strategy.

This 'Bungalyte' also used a power unit with motorcycle connotations in the form of a Norman type 'D', a 150 cc unit that first saw service in the Kenilworth motor scooter of the early 1920s. In this instance the Arthur Lyon Company of Victoria Street, London, supplied the electrics. A similar plant called the 'Focuslite' was intended to provide power for a portable cinema set.

Left: *The Chicago company Aermotor was a big-time manufacturer of all-steel windmills from the late 1880s. But windmills cannot provide power without a supply of wind, so Aermotor entered the engine-manufacturing market around 1906 in order to supply a cheap form of back-up power. In around 1909 a neat little air-cooled petrol engine was introduced, its salient feature being its rare mode of operation – eight-stroke. Although it looks complicated this Aermotor is actually quite simple in operation, with fewer power strokes leading to greater economy.*

Right: *Although once numerous, engines manufactured by Hamworthy, based in Poole, Dorset, are surprisingly rare in the preservation movement. The company was started in 1911 by Percy Hall and entered the stationary-engine market after Percy's brother Sidney joined him a couple of years later from Petter Limited. Sidney, known at Petter as Zacharia Hall, had previously been chief designer at Yeovil and was credited with introducing the two-stroke hot-bulb range that won the Grand Prix at the Turin show in 1911. The similarity between the two makes is obvious. This 4 hp Hamworthy is of 1931 vintage.*

Left: *The vertical air-cooled International Famous was built in 2 and 3 hp sizes from 1908. The general design resembled the better-known Famous and Victor water-cooled versions, but this model is lighter and was intended for portable applications where the saving of weight was a priority. A belt-driven fan is used to cool the cylinder in the normal manner. Cheap imports from the International Harvester Company and the Associated Manufacturers' Company represented direct competition for British-based engine manufacturers, many of which were struggling after bouts of austerity.*

Badge engineering

Anyone making a careful study of old stationary engines will soon become aware that badge engineering was rife, presenting a trap for the newcomer. Certain engine manufacturers produced batches of engines for equipment wholesalers, who in turn added their own name, brand name or logo. In some instances – presumably if the contract was large enough – the wholesaler's name was added to the larger engine castings such as the water hopper or the main body. In this way the true identity of an engine was instantly disguised and in some instances, with the passing of time, was lost altogether.

Just because an engine displays a name or logo does not necessarily mean that it was actually made by the manufacturer named. The term used to describe this method of marketing is 'badge engineering'. This Brownwall engine, built in the United States, was sold in the United Kingdom under the Melco (Melotte Separator Sales Company Limited) name but R. A. Lister & Company's Canadian branch badged it 'Lister'.

Another case of badge engineering, this time designed to make an engine more appealing to the Australian market, is seen with this 7 hp petrol engine made by E. H. Bentall & Company. The Essex-based company conjured up the name 'The Kangaroo' and added it to the crankcase inspection door. This company was a major exporter, shipping vast quantities of farm implements and machines to the colonies and Commonwealth.

Left: *The enclosed crank 'British Amanco' is another typical example of badge engineering. Made by Bradford Gas Engine Company Limited from mid 1928, it was a well-made engine built with longevity in mind, perfect for the typical Associated Manufacturers' customer. It was rated at 2½ hp at 500 rpm although, if pushed higher, 2.9 hp could be obtained. Governing was by the throttle method, using a butterfly valve in the carburettor, while a Wico EK magneto supplied the spark. A further advancement was in the use of ball-bearings to support the crankshaft.*

Right: *Bradford also supplied engines to Canavon Partners Limited. This Avonmouth-based company specialised in mechanically powered wood saws that it imported from the United States. When Second World War import restrictions curtailed this operation Canavon manufactured its own drag-saw while purchasing a suitable power unit from Bradford Gas Engine Company. This late example, built in 1946, still uses the older method of sight-feed lubrication.*

A typical modification that is so correct that it must have been carried out by a Lister agent during the engine's working life. This 1913 Lister 'L' is a very early example but the detachable cylinder head and later-style vaporiser, although correctly paired, are not in keeping with the age of this engine.

Although this engine might be mistaken for a variant of the popular and ubiquitous Lister 'D' type, it is in fact another confusing product of the Bradford Gas Engine Company's factory in Shipley, Yorkshire. The 2 hp 'King of All' was intended to compete with Lister's successful 'D' type but failed to do so because of its inferior build quality. These engines were painted green if sold under the Bradford guise and red (and marked 'AMANCO') when marketed by the Associated Manufacturers' Company in London.

Another trap for the unwary manifests itself in the form of engines that have been modified, updated or improved during their working life. Typical examples are period improvements comprising changes from LT (low tension) to HT (high tension) ignition systems and from paraffin to petrol engines, or vice versa. Some conversions are so original in appearance that the owner refuses to believe that the engine concerned is not as it left the factory. To add to the confusion, while some engines end up looking older, others look decidedly younger, while a third group consists of a mishmash of components from all ages.

The enthusiast studying stationary engines will come across a number of intriguing modifications, the origins of which have been lost over the years. This engine is a Lister 'L' type converted from tank to hopper cooling by means of an auxiliary water hopper bolted to the top of the cylinder head. The hopper is cast in two parts, each carrying its own part number, and was probably made by Lister's customer Millars' Machinery Company Limited. The latter in turn supplied the engine to Fletton Brick Company with a brick/stone crusher.

Pre-1960 diesel engines

Since around the mid 1990s small diesel engines have become quite collectable, and rightly so! Whereas later diesel engines all look similar in design, tending to lack character and collector appeal, this is not so with the examples of the 1930s, 1940s and 1950s. Engines built in this period carry their own individuality, with each maker claiming at least one unique feature. Although later products may feature aspects of modern technology, they tend to be boring and so are not as appealing to collectors.

Early small diesel engines, with an output of below 12 hp, offer an excellent opportunity to collect a range of engines that has yet to catch on fully among the older enthusiasts. Furthermore, certain makes, models or types of diesel engine

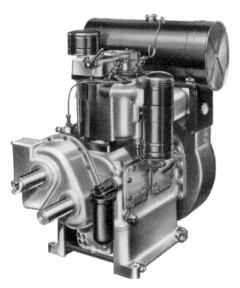

Left: *When Armstrong Siddeley Motors Limited of Coventry entered the oil-engine market in 1946 the company did so with an extensive background of automotive and aero engineering. The resulting range of engines were, if anything, over-engineered and built to last a lifetime – yet examples are few and far between in the preservation movement. The example seen here is the '14/20', a designation derived from performance output figures – 14 hp at 1000 rpm or 20 hp at 1500 rpm. Production came to an end when Hawker Siddeley, owner of the company, also acquired the Brush Group Limited, bringing competitors Lister and Petter into the same group of manufacturers.*

Right: *The Petter AVA1 is an air-cooled variant of the water-cooled AV1. Air-cooling is provided by a combined flywheel fan, which forces air over the finned cylinder and around the head, the air being distributed (in the normal manner) by a sheet-metal cowling. Series 1 AV engines were short-lived because of combustion-chamber problems and an indirect injection system that did not suit the engine. A makeover arrived in the form of a much-improved series 2, which used direct injection. The AV was introduced in 1947/8, followed by an air-cooled AVA in 1951. Seen here is an experimental version of the AVA being tested at the factory in 1948.*

are extremely rare and well worth obtaining, should an opportunity present itself. An added incentive is the chance to pick up a complete redundant lighting plant, such as a Lister 'Start-O-Matic', which could be restored before being put back into service – either for fun or for use during times of power failure.

There are many types of such diesel engine; those mentioned

The Petter PAZ1 first appeared at the Royal Agricultural Show at Blackpool in July 1953 and was claimed to be the world's first 1½ hp diesel engine of a commercial production type. It was built in single-cylinder form and produces 1½ hp at 1000 rpm, 2½ hp at 1500 rpm or 3 hp at 1800 rpm. A PAZ1T appeared in 1956 with a bell-housing for traction purposes. The size and weight of the PAZ1 makes it ideal for an enthusiast seeking a diesel engine to restore.

Below: *The Gardner 1L2, introduced in 1930 and rated at 11 bhp at 1100 rpm, is one of the best-known diesel engines in terms of performance and longevity. The ex-laboratory version shown here was built and supplied to Bristol College of Science and Technology in 1953 and transferred to the University of Bath in 1970. Laboratory sets built to special order by Gardner were usually powered by a 16 bhp 1L2 and direct-coupled to a Heenan & Froude dynamometer.*

Above: *The series of so-called 'Atomic' diesel engines was introduced by Petter Limited in 1928, with a range of sizes being extended both upwards and downwards in the early 1930s. The engines are of the two-stroke compression-ignition type, with the smaller sizes starting directly from cold with the aid of a compression release valve. Although tank cooling was favoured, a variation of hopper, spray or radiator systems was available to special order. The Atomic was built in sizes from 5 to 540 bhp in one-to four-cylinder configurations. Production ceased in 1940.*

Anyone interested in collecting diesel engines will have no trouble in acquiring a Lister 'CS' diesel. Examples such as the rusty but complete '5/1' seen here were built in large numbers for export and the home market. Many were supplied during the 1950s and 1960s with 'Start-O-Matic' lighting sets, some of which remain in service.

here are the most likely to be seen at engine rallies or in collections. Undoubtedly the most common diesel engine of the older style that you are likely to find is the Lister 'CS', of which thousands were supplied either to pump water or to produce electricity by means of a generating plant. The most common sizes are the 5 or 6 hp, known as the '5/1' or '6/1' 'CS' type, derived from five or six horsepower and one cylinder. A twin-

Initially known as the 'AD1' for a short time when introduced in 1954, the LD1 was intended as Lister's diesel replacement for the 'D' type, the production of which was coming to an end. Although the LD1 had a decent production run of over two hundred thousand, it was handicapped in terms of speed and output and was superseded by the LR in 1965 and finally by the LT in 1970. While a few LDs still remain in service, others are making their way into enthusiasts' collections.

Fowler of Leeds entered the diesel-engine market in the late 1930s with a range of engines designed by Arthur Freeman-Sanders, who was at that time a director of the company. The range, commencing with the 1D seen here, was comprehensive and unusual, covering every cylinder configuration from 1 to 6 with the exception of a 5! Although production is thought to have exceeded ten thousand, Fowler diesel engines are by no means common in the preservation movement.

cylinder 10 hp would be called a '10/2'.

Not so common but just as interesting is the Petter Atomic, first introduced in 1928 with the range of sizes being extended both upwards and downwards in the early 1930s. The Atomic was built in sizes from 5 to 540 hp in one- to four-cylinder form.

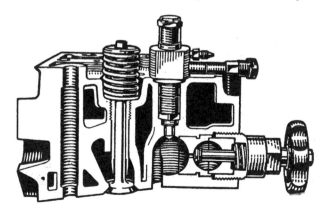

Harry Ricardo supplied the basic design of the popular Lister 'CS' diesel but the salient feature is its Freeman Sanders-designed variable-compression device, which ingeniously overcame the cold-starting problem so often encountered by smaller diesel engines. The main combustion chamber is connected via a small orifice to a secondary chamber, and by rotating a hand wheel positioned at the side of the cylinder head a valve is moved forward until it closes the orifice at the end of the movement. Consequently the decrease in clearance volume raises the compression pressure enough to give an instant cold start. As soon as the engine starts, the hand-wheel valve is screwed out to lower the compression to the working requirement – about 400 pounds per square inch.

Left: *Russell Newbery & Company Limited started manufacturing petrol and paraffin engines in 1909 but in 1930, following the general trend of engine manufacturers, switched production to diesel engines. The rare type RD1, illustrated here, was last used to drive barn machinery via line-shafting. Special features of 'RN' engines include horizontally opposed valves, which can be removed for inspection without taking off the cylinder head.*

Right: *Something a little different in the form of a French diesel engine built by CLM (Compagnie Lilleoise de Moteurs). Here we have an unusual opposed-piston configuration, the combustion chamber being halfway down the cylinder. This clever opposed-piston two-stroke design of Hugo Junkers was licensed out to numerous companies – it was taken up in the United Kingdom by Armstrong Whitworth and Peter Brotherhood Limited.*

Production ceased in 1940. Later, again from Petter, came the 'AV' series, initially with indirect injection (these are quite rare) and soon switching to the revised 'second series', this type using direct injection.

Other engines you will certainly come across include Fowler, Ruston & Hornsby, Russell Newbery and Bamford.

In the 1920s the transportation and assembly on site of large stationary engines was a slow and laborious task, yet engines were despatched to all parts of the world. This 12H 100 hp Ruston & Hornsby engine arrived at a railway station during the general strike of 1926 and the recipient was told, 'If you want it, send your own men and your own transport and use a 5 ton crane.' Once installed, it worked from 1926 until 1963, approximately fifty hours per week, driving plant in a leather works. The lorry is a Pierce-Arrow from the United States.

Large-power engines

As we have covered smaller, fractional horsepower engines, it is appropriate to look at a few examples at the other end of the scale. The first Diesel engine was made at the Augsburg Works, Germany, known to most people as 'MAN'. Mirrlees Watson & Company, Glasgow, manufactured the first British-built Diesel

Large engines take some restoring, especially if they have been vandalised to the extent of this 1913 Petter type 'GG' found in a forest in Norway. It had been the source of power for a turf factory working until the end of the Second World War. It will be a mammoth task to restore this extremely rare engine but the effort will be well worth it.

This picture of a Fairbanks Morse 'R' type does not do justice to the sheer size of the engine – the crankshaft and flywheels, as an assembly, weigh 4 tons! Development of the 'R' type began as far back as the late 1890s although the introduction date is listed as 1902. Built around 1906, this hefty example develops 50 hp at a meagre 300 rpm.

engine in 1897. It developed a meagre 20 hp but set a precedent for the large multi-cylinder open-frame giants that would follow. Huge slow-running, high-output, multi-cylinder engines with air-blast, and later 'solid' injection, evolved to usurp the steam engine's role as provider of power to drive factories or generating plants. Unfortunately most of the early powerhouses containing banks of multi-cylinder diesel engines of the old 'A' frame construction style have long since disappeared, leaving us only with a

Below: *Petter 'Atomic' two-stroke cold-starting heavy oil engines, such as this three-cylinder version, are now a rare sight, particularly at shows, because of exorbitant transport costs. Petter stated that over 50 per cent of these engines were sold for generating electricity because of their smooth running, low cyclic variation and low running costs. This photograph dates back to the mid 1970s, when this engine was seen at the Dorset Steam Fair; when it burst into life the sound resonated across the field, never to be forgotten!*

A Vickers Petter. The Ipswich factory in which these engines were built was constructed in 1913 for Consolidated Diesel Engine Manufacturers Limited and was equipped to make engines of up to a considerable 1000 hp per cylinder. Vickers Limited purchased the factory in 1915 in order to produce a new line of diesel engines for submarines. However, upon the cessation of hostilities in 1918 Vickers – rather like Westland Aircraft, Hewlett & Blondeau and so many other firms engaged in war work – found itself with empty order books, and the possibility of factory closure was looming. As a friendship already existed between Ernest Petter and the directors of Vickers it seemed a sensible solution to pool resources by moving large-engine production from Yeovil to Ipswich. A new company, Vickers Petter Limited, was formed in August 1919 in order to manufacture a range of large semi-diesel engines at Ipswich, leaving Yeovil to concentrate on small-engine production. This Foden mounted twin-cylinder Vickers Petter with Lister 'CS' donkey engine is one of the few remaining examples in preservation.

few contemporary photographs to remind us of those exciting times. The first high-power gas engines also date back to 1897–1900. These large installations, of between 500 and 8000 hp, were introduced to burn some of the blast-furnace gas, of which an over-supply was being produced and wasted. However, these giants, interesting as they are, fall outside the scope of this book; illustrated here is a selection of smaller 'large' engines such as you may come across in private collections or museums.

Further reading

Edgington & Hudson, *Stationary Engines for the Enthusiast*, first published in 1981, now in its 6th reprint.
Edgington, David W. *Amanco Engines*. Published by the author, 1996.
Edgington, David W. *The Lister D Story*. Published by the author, 2004.
Edgington, David W. *The Lister CS Story*. Published by the author, 2006.
Edgington, David W. *The Lister A & B Story*. Published by the author, 2008.
Website: www.stationaryenginebooks.co.uk. Or available, mail orders only, Vintage Reprint Services, Lodge Wood Farm, Hawkeridge, Westbury, Wiltshire BA13 4LA.

A–Z of British Stationary Engines. Kelsey Publishing Limited, 1996.
Stationary Engine Magazine. Kelsey Publishing Limited, monthly from 1974 to date. Website: www.stationaryengine.com

A Bristol Wagon & Carriage Works Victoria engine.

Places to visit

There are a number of excellent private collections of stationary engines that are open to the public on certain days each year. These are highly recommended as they are run by knowledgeable and enthusiastic people who have personally restored and operated the engines. Check in *Stationary Engine Magazine* (Kelsey Publishing Limited, Cudham Tithe Barn, Berry's Hill, Cudham, Kent TN16 3AG; website: www.kelsey.co.uk) for opening times.

Others include:

Amberley Working Museum, Houghton Bridge, Amberley, Arundel, West Sussex BN18 9LT. Telephone: 01798 831370. Website: www.amberleymuseum.co.uk

Anson Museum, Anson Road, Higher Poynton, near Stockport, Cheshire SK12 1TD. Telephone: 01625 874426. Website: www.enginemuseum.org (Includes a vast collection of engines that are restored and operated on site.)

Bradford Industrial Museum, Moorside Mills, Moorside Road, Eccleshill, Bradford, West Yorkshire BD2 3HP. Telephone: 01274 435900. Website: www.bradford.gov.uk

Dingles Steam Village, Milford, Lifton, Devon PL16 0AT. Telephone: 01566 783425. Website: www.dinglesteam.co.uk

Internal Fire (Museum of Power), Castell Pridd, Tanygroes, Ceredigion, Wales SA43 2JS. Telephone: 01239 811212. Website: www.internalfire.com

Lakeland Motor Museum, Holker Hall and Gardens, Cark-in-Cartmel, near Grange-over-Sands, Cumbria LA11 7PL. Telephone: 01539 558328. Website: www.holker-hall.co.uk

Museum of English Rural Life, University of Reading, Redlands Road, Reading, Berkshire RG1 5EX. Telephone: 0118 378 8660. Website: www.reading.ac.uk/merl

Science Museum, Exhibition Road, South Kensington, London SW7 2DD. Telephone: 0870 870 4868. Website: www.sciencemuseum.org.uk

Register of clubs and societies

There are around 150 relevant clubs in the British Isles, so wherever you live you will not have to travel far to be among friends with a common interest. The clubs are either dedicated totally to stationary engines or are part of a general group covering most aspects of vintage transport and machinery. A full list of clubs and societies is published annually in *Stationary Engine Magazine*.

The first engine introduced by Bamfords Limited was exhibited at Darlington Royal Show in 1920. It was similar in appearance to this rather robust-looking 6 hp horizontal engine. Initially a high-tension Hills magneto was tried but when this proved unreliable a switch was made to the Webster low-tension device (seen here). When the latter also gave problems in service a further change reverted to the more dependable high-tension system, this time via a Wico EK magneto.

Index

Printed and bound by CPI Group (UK) Ltd, Croydon, CR0 4YY

11/10/2024

01043560-0003